# LORRIES IN PICTURES 1965 - 2025

## CELEBRATING 60 YEARS OF THE COMMERCIAL VEHICLE & ROAD TRANSPORT CLUB

Peter Davies

HUGE JAM PUBLISHING

Published in 2025 by Huge Jam Publishing,
Gravenhurst, Bedfordshire.
www.hugejam.com

ISBN: 978-1-916604-60-5



Design, typography and layout: Peter J Davies



*Cover Photograph*: This 1965 Foden DAE6/32 of Glasgow based bulk liquids specialist James Hemphill Ltd was photographed on Beattock Summit just a few months after the CVRTC was founded. It took its power from an FD6 Mk.VI two-stroke and has the S24 cab - deemed to be the first tilt cab to be offered in the UK. It represents the sixties era perfectly and has all the hallmarks of the British classics that appealed to enthusiasts.

# PREFACE

The idea of producing a book to mark the sixtieth anniversary of the CVRTC was put to me some months ago by the committee as it seemed an appropriate way to celebrate such an important landmark. As one of the few founder members still here to tell the tale, I was more than happy to oblige.

The Club has been a big part of my life so I'm only too pleased to delve into my archives and dig out a representative range of pictures spanning the whole sixty years from 1965 to the present day.

Some of the shots I've chosen might have been published before, but many of them are new shots that have not previously been seen. It's been an interesting and rewarding project putting this book together and I hope readers will enjoy the content. So much has happened in these past sixty years and it's been a fascinating and very worthwhile exercise getting it all together.

It is sobering to think that back in 1965 no one had heard of a mobile phone or a digital camera. The internet didn't exist and there was no such thing as social media. How things have changed since those early days!

I dedicate this book to my dear wife Jill who not only put up with me messing about with lorries but gave so much of her time to help run the club. She was treasurer for many years and later, for a while, took over from Barbara Wimbush as Membership Secretary. On top of that she worked long and hard in dealing with all the paperwork for the early Classic Commercial Motor Shows, dealing with entry forms, issuing passes as well as being in attendance at the events to physically hand out plaques and fuel allowances. When I met Jill she was already involved to some degree in the lorry world since she was employed as secretary to the editor of Bedford Transport Magazine. The picture above harks back to happy times, some thirty years ago, where we're both suitably togged up for the occasion of the CCMS at Gaydon.

# The CVRTC - Sixty Years On

Sixty years! Yes, that's how long the CVRTC has been serving lorry enthusiasts. Whether you're a driver, an operator, an owner-driver or just anyone interested in road transport, the CVRTC has certainly got something to offer. Our members enjoy sharing their experiences as well as exchanging fleet information and photos, old and new. Over six decades the Club has brought enthusiasts together, both in the UK and abroad, and continues to offer a unique service.

So, how did it all begin? Before the CVRTC was formed there were numerous lorry enthusiasts who corresponded with each other about their hobby but nobody had attempted to form a nationally based club. There were clubs for those interested in pre-war commercials, mainly buses and emergency vehicles, but nothing for people interested in road haulage and lorries of all ages.

A gentleman called Norman Painting, who is interested in old Morris Commercials, had a letter published in the 1965 trade press suggesting a club for lorry enthusiasts. That triggered the idea amongst existing enthusiasts of bringing everyone to-gether in a properly organised club. In a short space of time a small band of some twenty like-minded friends, the author included, set about forming such a club and we decided to call it the 'Commercial Vehicle & Road Transport Club'.

We formed a committee of sorts in 1965 and assigned responsibilities to each committee member. We were all on a learning curve but, in due course, we began recruiting new members from all over the country. It's worth remembering that this was long before the days of social media and the internet – all communications were by post or 'landline' telephone.

The early days — this group shot of CVRTC members dates from 1969. L-R: Adrian Cypher, Mike Houle, Chris Salaman, Jill & Pete Davies, Denis Miller and Dave Sanders at the MMB in Thatcham.

Our early members included well-known lorry photographer Arthur Ingram, Chris Salaman who once worked for BRS, Denis Miller who was keen on showmen's transport, Steve Wimbush who worked at Western Express ECC, BRS historian Mike Houle, Adrian Cypher, Dave Sanders, Alan Wilson, Peter Seaword and Dave Rundle. We went on to welcome Raymond Jenkins, Colin Knapp, Glen McBirnie, Gordy Hayes, Marcus Lester, Clive Davis and Martin Phippard.

Other early members who spring to mind include Roger Kenney, 'Uggy' Newport, Mick Cook, Carl Johnson, John Mollett, Phil Moth, Tom Caren, Andy Burton, Len & Bob Neild, Pete Smith, Alan Paulton, Jim Read, Richard Storey, Nicholas Watts, Alan Pepper, Neil & Maggie Matlock, Richard Tew, Ron Wilkins, Danny Gedge, Andy Ping, Paul Redfearn, Robin Masters, Ray Stevens, Richard Cresswell, Bob Rust, Dave Stewart, Pete Sumpter, John Harrington, Trevor Stringer, Alan Dean, Alan Drake, Dave Mortimer, Dave Lee, Dave Norton, David Parry, Nick Eagle, Nigel Scaife, Martin Perry, Keith French, Rob Young, Timothy Byrne and Sid Duffin.

There are many other members who deserve mention if only space permitted so apologies to anyone not mentioned. Members include well-known names in the haulage business like Ray Hingley, Geoff Hinchliffe and John Pearson. Past members have included some high-ranking figures in the industry like Fred Hope, boss of a large haulage company and inventor of the Hope Anti-Jack-Knife device. Also Commercial Motor journalist Ron Cater who later joined Volvo Trucks at Irvine; BRS training manager Dave Tarbuck and, perhaps most famous of all, Frank Manton who founded Motor Traction Ltd, builders of Rutland Lorries.

As the Club grew we established a number of regional branches throughout the UK but, over the years, some branches have closed for one reason or another. Despite that, the Club has remained successful and well supported and still enjoys a nationwide following.

Sadly, we have said goodbye to some of our loyal members over the years but that is in-evitable. We are still recruiting new members and our Facebook page, ably managed by Richard Stanier, has a huge following. The digital age has transformed how we communicate and it's important to move with the times. However much we embrace the digital age there's still something to be said for a printed magazine so we plan to continue with our popular CVRTC News, edited by Graham Edge.

Looking back at past achievements, we can be proud of having created the Classic Commercial Motor Show, originally held at the Commercial Motor Lorry Driver of the Year finals at Cranfield in 1979. This grew into one of the UK's top annual events for lorry enthusiasts. Thanks must go to Alan and Sheila Wilson for their hard work in processing entry forms etc. once I and my late wife Jill had to pull out of organising the show owing to other commitments.

Our regular winter meetings held at the Old Crown in Alderman's Green, Coventry are greatly enjoyed. Members and invited speakers volunteer to do slide presentations on their favourite fleets and so on. Organiser, Club chairman Malcolm Mortimer, always makes members welcome as do the pub staff, while John Pearson of nearby Pearson's Transport plays an important role in securing the venue and welcoming members. There's always a raffle and an excellent free buffet to enjoy.

# The UK Haulage Scene 1965 - 2025

Over the six decades that the CVRTC has been in existence there have been enormous changes in the UK haulage industry and in vehicle manufacturing. These changes have largely come about through new legislation. We have witnessed the introduction of HGV drivers' licensing and tachographs; the impact of imported vehicles on British manufacturing; the growth in UK-European and Middle-Eastern traffic; the effects of containerisation; the decline of heavy industry; takeovers and group formation; intermodal traffic schemes; increasingly stringent exhaust emission controls; alternative fuels; the use of telematics and vehicle tracking - you name it!

Back in 1965, when the CVRTC was founded, hauliers were still adjusting to new legislation that had taken effect the previous year. Amendments to the Construction & Use regulations in 1964 introduced heavier vehicles, notably 32-ton artics.

*In the 1964 C&U regulations, ridiculously long outer axle spreads for the heaviest 8-wheelers made them unstable when tipping on site as well as difficult to manoeuvre.*

More legislative changes came in 1968 with the reintroduction of HGV driver licensing aimed at improving safety. Another safety-related rule was the requirement for annual MoT testing of HGVs. One good decision made was to increase the artic length limit to 15m. This enabled the carriage of 12m containers and also meant that 32 tons became legal on 4-axles - previously impossible with the quoted minimum outer axle spread.

The minimum outer axle spread for the heaviest rigid eights was also reduced, to 6.3m (20ft. 8in) in 1972, while the maximum gross weight was raised from 28 to 30 tons. This brought about a revival in eight-wheeler sales which had declined dramatically in the late '60s.

During the first decade of the CVRTC the UK commercial vehicle industry was very different from how it is now. In 1965 there were around twenty manufacturers but, by the mid-70s, at least half of them had either closed down or been absorbed into rival manufacturers.

British Leyland was responsible for killing off a number of long-standing makes like Albion and BMC (including Austin and Morris) while AEC, Guy and Scammell were to be next on BL's hit list. In 1970 Seddon absorbed Atkinson although the two makes were still marketed separately until 1975. By then Commer and Karrier were no more, becoming Dodge. Meanwhile, European imports were rapidly increasing with the likes of Volvo, Scania, Mercedes, Magirus-Deutz, MAN, DAF, Saviem and Iveco all keen to get 'a slice of the cake'.

UK-European trade was, of course, on the increase once Britain joined the EEC in 1973. Trouble was, the poor old British manufacturers had been put at a distinct disadvantage. While some bravely fought on, the outlook was becoming bleaker as we entered the '80s.

When it came to haulage companies, they also saw big changes. In 1965 British Road Services was still going but, by then, it was a limited company. The original state-owned British Transport Commission had been replaced by a new body called the Transport Holding Company in 1963 and BRS was reorganised, adopting a new fleet numbering system while the old door crest disappeared.

The Labour Government, under Harold Wilson, was in power from 1964 to 1970 and, as the original architects of BRS, they tried with limited success to expand state ownership through what was referred to as 'back door' nationalisation.

*Labour's attempts, under Harold Wilson, to expand BRS through 'Back Door' nationalisation led to a number of private hauliers, like George Read, losing their independence.*

The Tory Government soon put a stop to that when they regained power in 1970. By then the THC had become the NFC (National Freight Corporation). BRS was broken up into eight regional companies in 1972: Southern BRS, Western BRS, North Western BRS, Eastern BRS, North Eastern BRS, Midlands BRS, Scottish RS and Morton's BRS, each company having its own livery. From the late '70s onwards BRS was progressively run down by the Tories. It was to focus more on Truck Rental while BRS Parcels was re-named Roadline. 1982 marked a turning point when it was subject to an Employee Buyout under CEO Peter Thompson. The eight regional companies were reduced to five – BRS Southern, BRS Western, BRS Midlands, BRS Northern and Scottish Road Services. In 1987 Roadline (once BRS Parcels) was rebranded Lynx. As the decline continued, the haulage operations were renamed 'Exel Logistics' by 1990. BRS Truck Rental was sold to Volvo in 1998 and in 2005, Exel was sold off to DHL. What is left of BRS is owned by Volvo.

One significant legislative change that was brought in by Labour before they left office was the abolition of the old A, B and C licensing system which was replaced with a single O (operator) licence. This had quite an impact and marked the biggest change since the Road & Rail Traffic Act of 1933.

*Under new Labour-led changes to the carriers' licensing system in 1970, the old A, B and C licences gave way to a single 'O' or Operator licence seen here on the right.*

As we moved into the '80s, UK lorry manufacturing continued to diminish while imports rose. AEC, Albion, Atkinson, Austin, BMC, Commer, Dennis, Guy, Morris and Seddon had all disappeared from the scene. Seddon Atkinson was part of International Harvester who also owned DAF. Foden was absorbed into Paccar of the USA.

In the latter half of the '80s there was more bad news when, in 1986, Bedford was axed by its American parent GM. AWD bravely attempted to carry on where Bedford left off but by 1992 production ceased and what was left of the once famous Bedford brand was sold off to Marshalls of Cambridge. That too soon folded.

*When Labour reintroduced HGV Driver Licensing in the late '60s trainees had to display these prominent L Plates.*

Then it was announced in 1987 that Leyland had been taken over by DAF. Likewise, Ford became part of Iveco. Foden had already been absorbed into Paccar. Things didn't get any better in the 1990s – ERF lost its independence, becoming part of Western Star in 1996. Then, in 2000, Western Star sold ERF to MAN. It was clear that the UK industry was rapidly dying on its feet.

As we entered the new millennium there were basically three British manufacturers left, all of whom were in foreign ownership. These were ERF, Foden and Seddon Atkinson. Sadly all three were to face closure. Foden and Seddon Atkinson ended production in 2006 and ERF just a year later in 2007.

Since that date Britain has been entirely dependent on foreign manufacturers. Even the armed forces rely on Germany for their trucks. Okay – DAFs are built in Leyland so some might argue that Britain still produces trucks but, compared with 60 years ago when the CVRTC was first established, the industry has completely collapsed.

More legislative changes came about in the early '80s, the most important of which saw the maximum legal gross weight for artics raised to 38 tonnes (37.4 tons) in 1983, bringing about a new breed of 3+3 axle combinations. Another important change had taken effect in January

1982 when the use of tachographs in vehicles over 3.5 tons became compulsory, taking over from the old log books. This change was prompted by EU regulations.

*Determined to dissolve BRS, the Tories broke the organisation into autonomous regional companies each with its own livery. This Guy Big J wears the SRS livery.*

Since 2008, under an EU directive aimed at improving road safety, HGV drivers require a CPC (Certificate of Professional Competence). While existing drivers automatically qualify, they are required to undergo classroom re-training totalling 35 hours every 5 years. Some see this as a waste of time and money but drivers who don't comply can be faced with a fine of up to £1000.

On the haulage front some big, well-known names came and went in the '80s, not least Smith of Maddiston, part of United Transport and one of the UK's biggest fleets which was to cease operations in 1984. Smith's parent United Transport was a big group encompassing firms like Bulwark Transport, Leonard J Stamp, Wynn's of Newport and Sunter's. United's bulk liquid fleets ended up in P&O Road Tanks. In 2002 P&O European Transport was taken over by Wincanton plc.

Another haulage firm that everyone came to know was rapidly growing, namely Eddie Stobart. In 1985 the fleet numbered just 26 but by the early '90s Stobart's were a huge set-up with over 300 trucks. They continued to expand and by the year 2000 the fleet had grown to over 2500. Since 2021 it has been part of Culina, a subsidiary of the German Muller organisation.

Among the other major haulage companies was the aforementioned Wincanton, once milk hauliers, who underwent rapid expansion into other sectors in the '90s and by 2020 were overseeing around 4000 vehicles and trailers. In 2024 it was announced that Wincanton had been bought by American giant GXO (formerly XPO).

An earlier XPO acquisition was the French Norbert Dentressangle in 2015. This might not seem relevant to UK transport history at first but Norbert had earlier taken over the huge TDG group which incorporated around 50 of the UK's most famous hauliers – including the likes of Charles Alexander of Aberdeen, Harris Road Services, Crow Carrying Co, Russell of Bathgate and McKelvie's.

The growing use of containers from the mid '60s onwards led to a decline in traditional dock activity in ports like London and Liverpool and a corresponding increase in traffic at container ports like Felixstowe and Southampton. Famous fleets specialising in container work, like Brain Haulage and Russell Davies grew rapidly in the '70s and '80s. Brain Haulage of West Thurrock was originally established in 1950 and became a major force in container work, closing in 1992.

Another big name in the container business was Russell Davies which was formed in 1974 by Glyn Davies, son of old-established Swansea-based haulier Russell Davies. His fleet grew rapidly to become one of the biggest in the UK. In 1995 he sold the business to Securicor. Having apparently pulled out of haulage, Glyn Davies went on to form another successful operation in the form of Hanbury Davies in 1999. This too grew into a big fleet and in 2008 was sold to Wincanton Logistics.

Perhaps the biggest and best known of container operations is Maritime Transport of Felixstowe. In the quarter of a century since it was established it has undergone phenomenal expansion. Boss John Williams, whose father is Tommy Williams of TDW Distribution based in Bridgend, South Wales, founded the company in 2001 after buying Maritime Haulage. At that time the fleet numbered 136 vehicles and had 5 depots. Within ten years this had grown to 847 vehicles across 10 depots. In 2014 Maritime took over Roadways Container Logistics and the fleet grew to 1400 over 24 depots.

Expansion continued and in 2020 Maritime acquired Wincanton Container Logistics bringing the fleet up to 2000 vehicles. In 2023 the firm entered into intermodal traffic and began showing interest in electric trucks. Since

September 2024 Maritime has been part of the Swiss MedLog company based in Geneva. MedLog, or 'Mediterranean Logistics' in full, is a division of MSC (Mediterranean Shipping Company).

*Having absorbed the Transport Development Group, which included some of the UK's most famous fleets, Norbert Dentressangle was taken over by XPO of the USA in 2015.*

Over the years and long before the CVRTC was founded, there have been 'group' haulage operations like Transport Services Ltd back in the '30s which went on to include famous firms like General Roadways, CD&T, and Kneller & Chandler. The Transport Association and the aforementioned TDG were other big groups.

The Jigsaw Group of 2003 included C M Downton, F Swain, Bartrum's, Currie European, Elddis, Sparks, Pollock, Long's, Nolan and P C Howard. Downton then went on to acquire the whole Jigsaw group. In 2018, Hong Kong based EmergeVest founded EV Cargo which included C M Downton, Allport Cargo, NFT and Palletforce.

Another group, formed in 2012, is Kinaxia whose members include Bay Freight, William Kirk, Cammack Transport, Foulger Transport, Lambert Bros, Panic Transport, H R Maiden, Mark Thompson, David Hathaway, Fresh Freight and AKW.

Currently one of the UK's largest and most successful haulage companies is Turners (Soham) Ltd with a fleet approaching 2000 trucks and 32 sites employing 3400 staff. Originally founded in 1930, the firm has, in recent years, expanded rapidly through multiple takeovers and has a huge portfolio of famous contract customers.

On the legislation front a major announcement came in February 2001 when the Government permitted the use of 44 tonne (43.3 ton) lorries on UK roads. '44 tonners' had been allowed to operate since 1994 but only on road-rail intermodal work. In 2004 two hauliers, Stan Robinson of Stafford and Dick Denby of Lincoln campaigned, without success, to allow the use of 'roadtrain' style doubles outfits grossing 60 tonnes and measuring up to 25m. In 2011 Longer Semi-Trailers (LSTs) up to 15.65m were put on trial with some 1800 being allocated around 150 selected hauliers. After exhaustive trials the use of LSTs was finally given the green light in 2023.

*Denby of Lincoln, owners of this Scania, fought alongside Stan Robinson of Stafford, to allow 60 tonne, 25 metre doubles outfits on UK roads but were unsuccessful.*

The Government's decision to allow 44 tonners was made on the basis that heavier lorries could reduce the number of lorries on the roads – thousands fewer while saving up to 60 million lorry miles. Such statistics must have been music to the ears of MPs and the multitude of other anti-lorry bigots.

One can't help feeling frustrated by people's attitude to road haulage. Without an efficient road haulage industry such people would go hungry, not to mention going without other essentials of modern life – fuel for their cars, beer in the pubs, food in the shops and restaurants, clothing, household goods, building materials, electrical goods, furniture – just about everything they need in life!

# The first decade

*Top left*: This 1965 AEC Mandator Mk.V from BRS Grangemouth is heading west out of Edinburgh with an early Sea-Land ISO container.
*Above*: 1966 Seddon 13:Four artic tanker of ICI paints powered by a Perkins 6.354.
*Left*: Seen at Kirkcaldy harbour is this 1967 Guy Big J8 of the Cement Marketing Co, manufacturers of Blue Circle Cement.
*Below left*: A familiar sight in the sixties were NAFTA (GB) tankers like this 1968 ERF. Originally a Soviet owned brand, NAFTA later became part of Q8 Petroleum.
*Below*: In the sixties, so-called 'fossil fuels' were the leading source of energy in manufacturing and power generation. This 1969 Leyland Octopus of Garratt & Hemphrey is laden with coke.

# from 1965 - 1974

*Above*: While most lorries were British in the late sixties, Europeans were growing in numbers. This 1970 Scania 110 of Thomas Black is seen in Sheffield.
*Top right*: Atkinson lost its independence in 1970, becoming part of Seddon.This 1971 Borderer is seen at John Raymond, Bridgend.
*Right*: Marmalade makers Robertsons ran a fleet of Bedford TKs like this 1972 KGL wearing their controversial 'Golly' livery.
*Below*: Once one of the UK's best-known fleets, Robson's of Carlisle was noted for its 'named' lorries like this Foden S80 'Border Europa' from 1973.
*Below right*: In the early '70s Leyland dropped the Albion name but still used Albion model names with the G cab like this 1974 'Reiver' of John Hinchcliff's.

24
RUSSELL of BATHGATE
BEDFORD
RUSSELL
BSX 691C

This busy dockside scene shows a classic 1965 Bedford TK of Russell of Bathgate.

# Lorries in Pictures 1965-2025

Miers Transport of Wolverhampton specialised in milk transport. This Foden eight-wheeled tanker from 1965 features the S24 tilt cab introduced in 1962. At that time Miers' fleet numbered 55, including some lorries on churn collection.

This 1965 AEC 'Ergomatic' Mandator with fridge trailer belonged to Daniel Stewart of Blairgowrie who were taken over by Christian Salvesen in 1968.

J&H of Peckham was once a leading London haulier with 120 vehicles like this 1965 Atkinson T3046X artic seen at their Markyate Depot. They became part of HTS in 1971.

# Lorries in Pictures 1965-2025

*Top left*: Royston bulk hauliers A R Pigg & Son ran this Leyland-engined Bedford KHA seen on the A505 in Luton.
*Above*: Dodge moved up the weight scale with the new Cummins V8 powered K950 artic grossing 24 tons. This outfit of E&J Davis from Birmingham dates from 1966.
*Left*: Although new legislation in 1964 allowed 32 ton artics at the expense of rigids, tipper firms like Ellesmere Sand & Gravel stuck with eight-wheelers.
*Below left*: A newcomer to the UK scene in 1964 was Mack from the USA. John Brown of Leighton Buzzard became an agent. This 1966 R-series joined Pentus Brown's fleet second-hand from P&S of Glasgow. Seen on the A5 in Dunstable.
*Below*: Burton's Biscuits from Blackpool, famous for their 'Wagon Wheels' ran this 1965 Perkins-engined Commer VB.

# Lorries in Pictures 1965-2025

Guy's highly successful Big J range was a big seller in the 1960s with major fleets like British Road Services Ltd and, as seen here, British Rail, operating large numbers. This 1967 Big J4T is loaded with steel from its Park Royal terminal

In the '60s one could still see 'proper' drawbar outfits like this 3LV cabbed ERF of Gallin & Will of Falkirk from 1967 seen heading down the M6 Motorway.

It was in 1967 that Volvos like this F86 artic of Allison's Freightlines came on the scene, imported in growing numbers by the newly formed Ailsa Trucks of Barrhead.

# Lorries in Pictures 1965-2025

*Top left*: While European imports were growing, most UK makers were still holding their own, not least Foden with their S36 like this '67 Albion Sugar 8-wheeler. *Above*: Another British make that was still in big demand was Atkinson. Long distance fleets like T&M Catto still chose them. This L2486C is seen in London. *Left*: Traditionally in the light to medium sector, Bedford moved into the 16 to 24 ton weight class in 1966/67 with its KM range. JWG887E was one of a number operated by J&A Smith of Maddiston. *Below left*: Another Swedish newcomer was the Scania Vabis LB76 which sold in modest numbers. Scania sales took off with the arrival of the LB110 and 80. *Below*: Good old British Stalwart Scammell was still flying the British flag with outfits like this 1968 Highwayman.

# Lorries in Pictures 1965-2025

There were still up to fifteen British makes on the market in the late '60s. Among them was Seddon who were building outfits like this 32:Four maximum capacity artic. This nicely painted example belonged to Fred J Wright of Selby.

One drawback of the Scammell Routeman II was its single drive bogie. The last ones like this example from 1969 had Albion double drive, heralding the 1969/70 Routeman III.

Atkinson remained faithful to its 'outside rad' throughout its life even when attempting to modernise its image. One such attempt was the 'Viewline' first announced in 1966.

*Top left*: One make that never made it big in the 32 ton market was Dennis. The Maxim with its V8 Cummins was short lived. Dennis then concentrated on municipal vehicles from 1972 onwards. *Above*: Sadly Leyland was about to kill off the Albion name so this 1970 Clydesdale was among the last of the line in 1970. *Left*: After artic weights rose to 32 tons in 1964, 6x4 tractor units became popular as it was difficult to get 32 tons on four axles. Such outfits, like this S40 cabbed Foden of Laporte looked impressive. *Below left*: In the lighter class Ford's D series was a popular choice and a rival to Bedford's TK. This 1970 model of Ryman Conran has a truly striking paint job. *Below*: Volvos, like this Dawsonfreight F88, were still novelties in 1970 but were soon to outnumber home-grown trucks.

# Lorries in Pictures 1965-2025

Twin-steer or 'Chinese Six' units were a phenomenon of the late '60s/early '70s, brought about by the 1964 C&U changes. While 32 tons was allowed on four axles, the original 13m length limit couldn't accommodate the minimum outer axle spread. Twin-steers were one answer as they were lighter than 6x4s and could operate with existing tandem axle trailers. Road tax was also lower. AEC's Mammoth Minor, like this Frigoscandia outfit run by Simpson's of Wisbech, were popular.

While 32 ton artics gained ground some operators still ran 8-wheelers which could also gross 32 tons with trailers. This striking ERF outfit belonged to DTD Farms, Redditch.

The Perkins V8 powered Mastiff, from Leyland's BMC range, grossed 28 tons. This outfit of Hodgkinson's Transport, North Hykeham, was on contract to Ley's Castings.

*Top Left*: Dennis bid for a share of 16-tonner sales with their 'Dominant' but it didn't have much impact. This lwb flat of Wood's from Ashton-under-Lyne dates from 1971. *Above*: The Guy Big J4T was a highly successful machine. It appeared in many well-known fleets including Ackworth Transport of Pontefract who owned this 1972 outfit. *Left*: Many Scottish operators were quick to switch to European imports in the 1970s. Well-known haulier, Hugh Clelland of Stepps ran this nicely liveried 1971 Scania LB110. *Below left*: Another manufacturer to build twin-steer units in volume was Fodens Ltd. This DAXB8/38 of Redland's has the S60 cab with forward-sloping screen. *Below*: From 1972 eight-wheelers could legally gross 30 tons. An excellent example is this ERF 68GXB of Barton-le-Cley based L G Bradshaw. It features the 8LV/L fibreglass cab.

Though Leyland was clearly determined to phase out the likes of Albion, AEC, Guy and Scammell, these once famous makes still survived into the early '70s. Alas this 1973 Reiver of Ayr based J Wilson & Son would have been among the last Albions to wear the badge. NAG154L is seen at Laporte Chemicals' Kingsway plant in Luton.

Scammell's highly successful Routeman III eight-wheeler with its Michelotti cab became a popular basis for fuel tankers like this example of Shell seen on the M4.

Volvo sales really took off with the arrival of the F88 and by 1973 when this outfit entered service with W H Bowke they were selling like the proverbial hot cakes.

*Top left*: Commer launched its new 'Commando' in 1974 and this box van of Creative Printers, based in Wellingborough, was among the first to enter service.
*Above*: By 1974 the Albion name had finally disappeared but some Albion model names lived on with the new G-cabbed 'Blue Line' models like this Mixconcrete Reiver mixer.
*Left*: ERF announced its A-series in 1970 designed by ex-Dodge man Alan Turner. It had a completely new chassis and 7LV cab. This 1975 Arrowsmith artic is a nice example.
*Below Left*: All the former medium-weight makers were moving into the heavyweight sector in the 1970s including Dodge, with its K3820P based on the Spanish Barreiros.
*Below*: Even Bedford entered the maximum weight artic sector with its 2-stroke Detroit Diesel powered KM 32 Tonner seen here at Aberthaw's Cement works near Cardiff.

# The second decade

*Top left*: Until the mid-seventies Seddon and Atkinson had kept their individual identities. This 1975 Cummins powered Seddon 34:Four wears the superb livery of Smith of Maddiston.
*Above*: Dating from 1976, this Scammell Crusader is in the livery of Baird's Malt but was operated by S J Beckwith of Chelmsford.
*Left*: This Dairy Crest Bedford TK1020 6-wheeler was factory built under SVOS8701 and grossed 15 tons. It was developed for customers wanting a low loading height. Dairy Crest was the MMB.
*Below left*: Another European make to appear in growing numbers during the '70s was DAF, this 1977 F2200 tanker of LPG Transport being a typical example.
*Below*: Fodens Ltd was still going strong in 1978 when this S83 livestock truck was built but just two years later it was taken over by Paccar.

# from 1975 - 1984

*Above*: The Leyland T45 'Roadtrain' marked the culmination of the company's rationalisation programme. It was their last new range before the DAF takeover. This is an early dealer demonstrator.
*Top right*: Rathcoole, Dublin-built Dennisons were relatively rare in the UK and BVT was one of the few firms to operate them.
*Right*: By the early '80s, when this 142 joined Russell Davies, Scanias had become huge sellers in the UK and have remained so.
*Below*: Like Foden, ERF was still a major player in the UK market in the '70s and early '80s. This smart bulk tanker of Earls Cement is a C-series 'C31' from 1982.
*Below right*: After Chrysler bought Rootes in 1967 the Commer name was progressively changed to Dodge as on this 1984 Commando 2.

# Lorries in Pictures 1965-2025

One of Leyland's big announcements of the decade was the new Marathon of 1973, developed by AEC. It was seen as a possible saviour for the loss-making company. Sadly British Leyland's fortunes continued to decline so the UK Government stepped in and nationalised it. Marathons were produced at Southhall and later at Guy's factory in Wolverhampton

Foden was always a strong contender in the UK rigid-eight sector and the S80, seen here, found its way into many leading tipper fleets including Selby's Garage.

Sadly, one of the UK's most famous makes, AEC, was to disappear in 1977 at the hands of BL. This 1975 5TGM Mercury is a late model lacking the famous blue triangle.

*Top left*: Having built its Detroit powered KM 32 tonner, Bedford launched its TM heavyweights in 1974. This TM2600 tanker was new to Corralls in 1975.
*Above*: ERF's B-series arrived on the scene in 1974 and was the first ERF with a tilt cab. The SP (Steel/Plastic) cab served as the basis of subsequent cabs up to the last ECX of 2002.
*Left*: Magirus Deutz was gaining ground in the UK in the mid '70s. This brand new 232 artic tanker was being delivered to United Molasses in Liverpool in November 1975.
*Below left*: Gone are the days, sadly, when Scammell dominated the UK heavy haulage scene. This Econofreight Contractor, ex-Wynns, is seen hauling a large transformer en-route to Sundon sub-station near Luton.
*Below*: Seddon Atkinson's first new lorry was the 400 series of 1975. This Schweppes artic is pictured in Edinburgh when new.

ERF's leading seller in the 1970s was its highly successful B-series which was offered with a choice of day cab or sleeper. This impressive artic tanker of Calor Transport from 1977 has the high-roof sleeper cab. Calor Transport was expanding in the seventies, acquiring several rival tanker fleets.

Fodens Ltd continued to maintain a strong presence in the rigid eight sector with models like this 30 ton gross S83 of Tarmac dating from 1978.

Having launched the TM3250 with the Detroit 6V-71 in 1974 Bedford added the TM3800 38-tonner in 1976 with the 296bhp 8V-71 and a full-width cab.

# Lorries in Pictures 1965-2025

*Top left*: A slump in sales after the 1964 C&U amendments led Leyland to suspend 8w production in 1970 but it later re-entered the market with the 500 series Octopus. This Redland tipper dates from 1978.
*Above*:Europeans like Scania were steadily increasing their share of the UK market in the '70s with trucks like this fabulous 1977 Scania 111 belonging to Swain's of Stretton.
*Left*: UK manufacturers like Seddon Atkinson, owned by International Harvester, bravely fought back with machines like this 400 series of Henry Nuttall from Heywood.
*Below left*: Another import to gain popularity was DAF who introduced rigid-8s to meet UK demand. This lwb flat of Goddard's dates from 1977.
*Below*: Like Bedford, Ford moved into the heavy sector in 1975 with its Cummins powered Transcontinental, built in Amsterdam. London hauliers Samuel Bros owned this 1978 artic.

# Lorries in Pictures 1965-2025

Ailing British Leyland soldiered on through the seventies with the Marathon which, by 1979 when this Rolls powered Marathon 2 joined the G L Baker fleet, was beginning to show its age. G L Baker's biggest customer was Marks & Spencer's foods and the firm became BOC Transhield before later evolving into today's Gist Ltd.

In the medium weight class Ford's long serving D-series was a leading seller. It scored points over rival Bedford in having a tilt cab. This Hanson box van dates from 1979.

From 1978 to 1982 British fans of American trucks were treated to the sight of White 'Road Commanders' like this artic of C S Palmer, formerly owned by Lock of Headcorn.

*Top left*: Seddon Atkinson was still a big seller throughout the '70s. This '400' 6x2 with tag axle was engaged on European work with F B Atkins, Derby. *Above*: A rare sight in the UK was the Irish Dennison, especially in 8-wheel form. This tipper began life as a demonstrator and later joined P&R Bushell in Halkyn, North Wales. *Left*: Another European make to sell quite well was Iveco's Fiat 170, like this artic of Welch's based in Witham. *Below left*: Dodge replaced its 500 series KT900 8-wheeler with this 30 ton C3084 based on the Spanish Barreiros. This tipper belonged to King's of Chelmsford. *Below*: A newcomer to the British scene was the Hino HE335E which began appearing in the '70s. They were assembled in Ireland by Harris Ltd of Dublin.

Bedford entered the eighties with a comprehensive model line up covering the whole weight range from 1 ton to 42 tons and was still enjoying healthy exports. Admittedly the TK was showing its age, having been in production for 20 years. Despite this it was still a big seller to fleets like Roadline, the new name for BRS Parcels. This TK1020 dates from 1981.

Scania continued to strengthen its position in the UK with new models like this 112m introduced in 1981. Malcolm Woodhouse is a leading bulk haulage firm from Lancaster.

As well as the T45 Roadtrain tractor units, Leyland built the new Constructor 8-wheelers which were developed by Scammell using their expertise from the Routeman III.

*Top left*: Foden lost its independence in 1980 when it was taken over by Paccar of America. It continued to be a leading UK manufacturer especially with 8-wheelers like this Longcliffe S108 bulk tanker.
*Above*: In 1980 Bedford finally offered a tilt cab in the form of its TL. Some joked that TL stood for 'too little, too late'. BRS Truck Rental owned a fleet of them.
*Left*: Volvo announced their important new F12 in 1977. By then they were turning out Volvos at their Irvine factory in Scotland. Turners of Soham were early customers.
*Below left*: British enthusiasts were treated, albeit briefly, to the sight of Kenworth K100s around 1979-1981, imported via Dando's of Chipping Sodbury.
*Below*: The Europeans were all gathering pace in the early '80s. The DAF 2300 was popular with firms like Bulwark seen here.

# Lorries in Pictures 1965-2025

From 1983 the UK artic weight limit rose to 38 tonnes (37.4 tons) and this gave rise to a new breed of truck like this 5-axle Scania 112 outfit of Adam Jones of Blackheath, Birmingham. Some units, especially European makes, had tag ax added as seen here while most British manufacturers developed 6x2 'rear-steer' units.

Most British fleets welcomed the new heavier artics, boosting sales for manufacturers. This MMB Dairy Crest outfit is hauled by an ERF C-series 'C41' with steering pusher axle.

As well as the standard C40 cab, Leyland's Roadtrain wa available with the high-datum sleeper version as seen on this 1983 artic of Midlands BRS.

*Top left*: The European manufacturers soon caught on to the importance of the 6x2 rear-steer units in the UK. This 1984 DAF 3300 of Perfect's from Gt. Yarmouth is proof.
*Above*: Likewise Mercedes, who had a lot of experience of such units, offered the 2028 which became an important seller. This outfit was on European work for Chambers & Cook.
*Left*: There was continuing demand for rigid eight-wheelers and Seddon Atkinson, the makers of this '301' lorry of Redland Roof Tiles, were major suppliers.
*Below left*: Another important European make to enter the scene was Iveco, formerly Fiat, whose 190-30 'Turbo' is seen here with an added 'pusher' axle in Dutton Transport's fleet.
*Below*: Drawbar outfits of this type were not that common but one firm to run them was Carman Brit-European of Stoke, the owners of of this underfloor engined MAN 19-321.

# The third decade

*Top left*: B-prefix registration marks as seen on this DAF 2800 of Robson's were in effect from August 1984 to July 31st 1985 so 'Border Sultan' qualifies for inclusion here.
*Above*: An impressive outfit by any standards is this 6-axle Mercedes 2033 of R J Roberts, Holyhead coupled to an Irish Stateline trailer.
*Left*: ERF's E10 with its Cummins LT10 was good seller in the'80s. Onward of Pontefract were big ERF customers. Though advertising vegetables, this artic was carrying waste paper.
*Below left*: Well-known Nottinghamshire firm Maurice Hill opted for this Iveco Turbo for long distance work, seen here on the M1.
*Below*: Collaboration with DAF and Pegaso under International Harvester led to the launch of Seddon Atkinson's new Strato in 1988 with the shared 'Cabtec' cab.

# from 1985 - 1994

*Above*: In the late '80s and early '90s ERF was about the only true British manufacturer left. Many loyal ERF customers stood by the Sandbach company, Brian Harris from Devon being one example.
*Top right*: Bedfordshire haulier H E Payne, now owned by WS, favoured Volvo and this superbly liveried F10 is typical of their 1990s fleet based at Wyboston.
*Right*: Another fabulous East Anglian fleet is that of Knowles of Wimblington, this 1991 Scania 113 being a perfect example.
*Below*: A slightly more unusual machine is this Tango liveried ERF ES6 drop-frame curtainsider for drink distribution from 1993.
*Below right*: As Leyland's fortunes declined DAF stepped in to form Leyland DAF. What was the Constructor became the '80 Series'

Roped and sheeted loads on flats were still fairly common forty years ago when this shot was taken at Robson's souther depot in Flitwick. By that stage the firm had been taken over by their main customer United Glass. Thankfully they retained Robson's distinctive livery and the practice of naming the vehicles, this Roadtrain being Border Ophelia.

As well as heavy artics MAN entered the 16 ton sector with the likes of their 16.170, seen here in the form of a TNT contract vehicle for Cow & Gate, now part of Danone.

Many well-known haulage firms, like Bassetts of Tittenso became firmly hooked on European makes in the '80s. Volvos, of course, like this FL10 were UK built at Irvine.

*Top left*: G W West & Son of Burton-on-Trent ran a number of Leyland Freighter four-wheelers including one with a twin-steer conversion.
*Above*: ERF's E-series of 1986 was more than a match for any European import, offering excellent comfort and all-round performance.
*Left*: London Carriers of Croydon, the 'in house' fleet of Philips Electronics, once ran Guy drawbar outfits but went on to Seddon Atkinson 301 artics in the '80s.
*Below left*: Mercedes-Benz were slow to offer an 8-wheeler but when they finally launched the 3025, seen here, it sold well.
*Below*: Nearly all the European makers came to acknowledge the importance of 8-wheelers in the UK and came up with suitable models. This Iveco 300-25 took over from the earlier 'Maggie' Deutz.

By the late '80s imports were growing in numbers and some of the UK's leading manufacturers had ceased production. The scene was rapidly changing with the likes of this Iveco Turbo replacing British makes like AEC, Guy and Bedford. Having said that outfits such as this one of T Brady from Barrow-in-Furness had an appeal all of their own.

Another European make to gain ground was Mercedes with trucks like this 1617 of Torridge Transport from Devon. It has a pusher axle added raising the gvw to around 20 tons.

At the lighter end of the market, Ford's Cargo outlived its Bedford TL rival but soon it was absorbed into Iveco. Production at Ford's Langley plant ceased in the '90s.

# Lorries in Pictures 1965-2025

*Top left*: ERF's E-series was a real winner wth its well-appointed sleeper cab and choice of power units. The E10-325, like this smart outfit of John Adams, Linlithgow, competed well with the foreign imports.
*Above*: Although Foden Trucks was foreign owned it was still viewed as British. This 1988 Paccar-inspired S106T 6x4 hauls an unusual mixer trailer with Morrison.
*Left*: Another Foden-Paccar development was the use of Caterpillar diesels. This 4300 tipper wears the livery of Coventry based Pearsons Transport. Boss, John Pear-son. plays an important role in the CVRTC.
*Below left*: A classic 38-tonne artic of the '80s is this superb Volvo F10 of H Whittaker. Note the tag axle as opposed to a 'pusher'.
*Below*: Old-established Northampton haulier Wreford's remained loyal to Seddon Atkinson with outfits like this 1988 Strato.

One of the most admired fleets in the West Country, and indeed in the whole country, was that of Brian Harris whose yard was at Bovey Tracey. His superbly liveried wagons stood out wherever they went. Sadly, all good things come to an end and the end came for the Brian Harris operations in 2001. This 1989 Seddon Atkinson Strato epitomises the famous fleet.

Another nicely liveried fleet is that of Maurice Hill who is thankfully still going strong. This Iveco 190-30 artic is seen delivering materials to a construction site in Flitwick.

While on the subject of liveries, here's another much admired fleet from 'North of the Border'. This smart 1989 ERF E14 wears the colours of J B McBean, Kirknewton.

# Lorries in Pictures 1965-2025

*Top left*: Once Bedford production ended in 1986, the Dunstable plant was taken over by AWD and some TL based AWDs like this Perkins powered 8-14 dropside continued to be built.
*Above*: Eddie Stobart rose to become one of the UK's most famous fleets. Early vehicles included Ford Cargo drawbar outfits on their Canco contract. In 2021 Stobart became part of Culina, a subsidiary of Muller.
*Left*: Ken Thomas of Guyhirn is another much admired fleet. This fabulous 1990 Scania 113M is typical of their vehicles.
*Below left*: After DAF took over the Leyland operations in 1987 the Leyland name was added to right hand drive DAFs as seen on this 95 of K Transport loaded with a steel coil.
*Below*: Blackburn based W H Bowker ran a number of these big volume drawbar Volvo 'Eurotrotters' on UK-European traffic.

Leyland DAF continued to market the former Constructor six and eight-wheelers badged as the '80 Series'. However they were not to last long and by 1993 the C40 cab was phased out. DAF added black stripes to the front panel which robbed them of valuable signwriting space. Builders' Merchants E H Smith ran this 1991 model with self-loading crane.

Scanias took on a new look in 1991 with the launch of the 3-series 'Streamline' seen here in the livery of Taylor's of Martley, Worcester. Taylors once ran tippers in the fifties.

Iveco Ford's new Cargo hit the headlines in October 1991 and soon found its way into some important fleets like Meachers, seen here. It was voted 1992 Truck of the Year

*Top Left*: With European trucks rapidly growing in numbers it was reassuring to see a few ERFs still doing their stuff. This E10 8-wheel tipper was working on re-surfacing in the centre of Flitwick.
*Above*: DAF rebranded the old Leyland Roadrunner '45 series', one example being this 1992 delivery truck of Graham's.
*Left*: No one could argue that the Iveco 'EuroTech' wasn't a good looking vehicle especially when painted up in a nice livery like that of Stalker's of Carlisle.
*Below left*: The Austrian Steyr was sold briefly in the UK in 1988-94. One firm to run them was PDM of Leighton Buzzard. PDM stood for Pat & Dorothy Moroney.
*Below*: A new arrival on the scene in 1990 was the mighty Renault Magnum, this Mack engined outfit was the pride and joy of owner driver Rod Izzard of Flitwick.

Hauling steel rail is left to specialist firms like Lawsons Haulage of Cockermouth who have extendable 'trombone' type trailers as seen here coupled to a beautifully liveried Foden S106T dating from 1993. Lawsons take great pride in their vehicles which carry fleet names, this Foden seen taking a break at Toddington M1 Services being 'Mountain Destroyer

All the European manufacturers came to acknowledge the UK's demand for rigid eights especially for tipping work. This 1993 F90 cabbed 32-322 of McNicholas also has a grab.

Welshman Glyn Davies formed Russell Davies in 1974 o containers, moving into distribution in 1988 before sellin to Securicor in 1995. This '93 Iveco is typical of the fleet

*Top left*: This traditionally liveried Leyland DAF 95-400 artic of H Whittaker from Macclesfield is a particularly handsome outfit dating from 1993.

*Above*: Famous Somerset haulier, R T Keedwell of Highbridge and its many subsidiaries operate throughout the UK. This Scania 113M Streamline is typical of their fleet in the '90s.

*Left*: Mercedes' leading model of the early '90s was the V8-powered SK seen here in the form of this 1850 artic of Highland Transport from Carrbridge near Inverness.

*Below left*: When the Renault Magnum was launched in 1990 it caused quite a sensation. It boasted the tallest and roomiest cab on the market. M500AVF wears the smart livery of J&C Finney of Tyldesley.

*Below*: Another memorable livery is that of Fenland bulk haulier Hankins of Manea, the owners of this 1993 ERF EC14.

# The fourth decade

*Top left*: Volvo's FH12 was well established ir the UK by 1995, the year that it also entered production at the Irvine factory in Scotland. Massey Wilcox of Cullompton was one of many fleets to run large numbers of them. *Above*: The Renault Magnum became a firm favourite with hauliers like Jordan Transport who operated services to Europe. *Left*: One of Scotland's finest fleets is D Steve & Son of Wick who have been going for over century. Their main traffic is seafood. This fabulous Scania 144L V8 dates from 1997. *Below left*: A&V Squires, from Southwell nea Newark, had this DAF 95XF new in 1998 to operate on heavy haulage up to 50 tons gcw. *Below*: Yet another nicely liveried fleet is that of Ruthin, North Wales, based L E Jones Inter national, the owners of this 1999 Actros 2540.

# from 1995 - 2004

*Above*: Though increasingly out-numbered by foreign imports, some UK manufacturers fought on, not least ERF with their highly successful EC like this EC11 of HPH, Preston.
*Top right*: Equally, the other Sandbach maker, Foden, was still very much in business albeit as a subsidiary of the American Paccar Co. Arclid Transport were once big Foden users.
*Right*: ERF had become part of Western Star in 1996, then it was sold to MAN in 2000. MAN themselves were selling well in the UK with their TGA like this outfit of Top Gear.
*Below*: Another UK maker to fall into foreign ownership was Seddon Atkinson. This 2003 Strato of F J Wynn was built at Pegaso's factory in Madrid. It had an Iveco cab.
*Below right*: Iveco unveiled its new Stralis in 2002. B J Waters were early customers.

# Lorries in Pictures 1965-2025

In their native Sweden Volvos were widely engaged on timber transport so this FH12-420 six-wheeler and drawbar trailer of John Weaver & Sons from Staunton-on-Arrow in Herefordshire would, no doubt, have felt quite at home! N469YCJ was caught on camera near Ross-on-Wye in April 1998.

ERF's big news in the early nineties was the launch of the new EC in 1993. They competed well with European makes. Intermodal Scottish haulier John G Russell ran a number.

While Seddon Atkinson remained in production for a ti under Iveco ownership they were progressively changed This 1995 Strato of Smiths of Scotter has an Iveco cab.

*Top left*: The new Mercedes Benz Actros was unveiled at the 1996 IAA Hanover Show and Bury based J&H Hinchliffe were among the earliest UK customers.
*Above*: This 1996 Foden 3335 artic of Gilbraith Tankers has the later Mk.5 S10 cab on which the company reinstated the famous kite motif that was missing on the Mk.4.
*Left*: Historically Scania had not been a serious contender in the 8-wheeler sector but models like this 1996 94C-310 of Gibbs & Dandy were highly successful machines.
*Below left*: Clearly designed for European work this lowered Leyland DAF 95 of Currie Group's Reliable Trucking meets the European 4m maximum height limit.
*Below*: Iveco sales were growing in the '90s with models like this 1996 high-roof Eurostar of IMP Freight Services proving popular for long distance and European operations.

# Lorries in Pictures 1965-2025

We all thought it was all over for Bedford when GM ceased truck production but the famous make saw a brief comebac following AWD's closure in 1992 and the transfer of assembly to Marshall SPV of Cambridge. GM gave Marshall the right to use the Bedford name. This 1996 TL17-21 of A E Costin had a Perkins Phaser 6-210ti and a Rockwell rear axle.

MAN 8x4 tippers continued to sell well during the nineties. This F2000 cabbed example of Keanes based in Stanmore is seen on the old A421 near the Brogborough landfill site.

LTS of Oldham's lorries stand out by virtue of their striking livery. The firm also provides heavy haulage services This Renault Premium is typical of their fleet in 1997.

*Top left*: F A Hawkins of Eaton Bray is a well-known Bedfordshire haulier who always ran ERFs. The firm will reach its centenary in 2026 and now runs DAFs. *Above*: Another large Bedfordshire firm is H E Payne, noted for its superb livery. They were taken over by WS Transportation in 2021. This Scania dates from '97. *Left*: Reminiscent of a scene from New Zealand perhaps but this Foden 4525 timber truck of KKM Haulage was photographed on the A66 in Cumbria. *Below left*: Elddis Transport is one of the UK's best known hauliers and is based in Consett, Co Durham. This 1998 Actros 2540 is typical of their late '90s fleet. *Below*: Swain's of Strood, the owners of this Volvo FH, ranks as one of the biggest hauliers in the UK with no fewer than 12 depots, 600 vehicles and 900 trailers.

# Lorries in Pictures 1965-2025

ERF's EC11 was selling well in 1999 when this bulk tanker joined Bowmans Flour in Hitchin. At that time ERF had be in Western Star ownership for three years. The EC was launched in 1993 and the leading power unit was the Cummins M11, launched around the same time. In 2000 ERF production moved from Sun Works, Sandbach to Middlewich.

Little did people think that ERF, Britain's last manufacturer, would be sold to MAN shortly after this TGA artic entered the fleet of bulk haulage specialists James Booth in Bolton.

Few trucks have captured the imagination of enthusiasts more than Scania's 'T cab' like this 144G outfit of AIG heavy haulage. Scania ceased T-cab production in 2005.

*Top left*: Scania's 4-series, launched in 1995, was still in production in 1999 when this 124L joined the fleet of Allied Mills. It has the lowest height P cab and the Euro 3, DSC12 engine with 400bhp.
*Above*: As the millennium dawned, Fodens were still selling well. This 1999 Alpha of J R Foster from Hopwood, Birmingham is seen on the A14 near Cambridge.
*Left*: Time was running out for Seddon Atkinson as we entered 2000. This Strato of Helston based Robert Benney resembles an Iveco EuroTech with a Big A badge.
*Below left*: Old established S J Barrick of Goxhill in North Lincolnshire specialise in bulk powders. This Mercedes Actros 2540 is seen near Stoke Ferry.
*Below*: Weaver Transport from Whitchurch in Shropshire operated this smart Volvo FH12 Globetrotter seen here on the A14.

As we entered the new millennium this nicely-liveried Mercedes Actros 2543 was a new addition to the fleet of Ken Abram based in Skelmersdale, Lancashire. The company was well-known and admired by enthusiasts. It provided warehousing and distribution mainly in the food, drink and paper industries. Sadly it entered receivership and closed in 2006.

Once MAN had bought ERF some models began to appear with adaptations of MAN cabs, like the ECS seen here in the form of this 2002 brick wagon of Corby Glen Transport.

From 2000 onwards ‘Leyland’ was dropped from Leyland DAF so all were badged DAF like this superbly liveried 95XF of Ken Thomas Ltd based in Guyhirn, Cambs.

*Top left*: The pictures on this page show a selection of some of the finest fleet liveries beginning with Edwards Transport of Market Drayton's superb Scania 144L V8 Topline seen on the M6 near Birmingham.
*Above*: Another excellent livery appears on this Volvo FH12 Globetrotter of E Pawson & Son Ltd from Rotherham.
*Left*: This 2001 Foden Alpha with high-roof 'XL' cab wears the superb livery of Lawsons of Cockermouth. The fleet name 'Mountain Destroyer' was transferred from their 1993 Foden 4380 tractor L622KHH.
*Below left*: J H Jones from Colwyn in North Wales were the proud owners of this Iveco EuroStar. The writing on the high roof sides reads 'One Step to Heaven'.
*Below*: Fords International, Derby is another nicely turned out fleet as seen with this 2002 Renault Magnum artic seen on the M1.

It's been apparent for some time that liveried fuel tankers from the likes of Shell, Esso, BP and Jet have been diminishin in numbers as if they don't wish to advertise the brand names. Perhaps pressure from the climate change lobby to switch to electric cars has influenced the change to plain tankers from contractors. This ERF ECT outfit of Shell dates from 20(

Once loyal to ERF and Leyland, Chamberlain Transport of Crewe switched to Scanias like this impressive 124L-470 Topline of 2003. Their excellent livery still lives on today.

Iveco unveiled its highly successful new Stralis range in 2002 and it soon found some high profile customers like W Tabern based in St Helens. This artic dates from 2004.

*Top left*: In 2004 two leading hauliers, Stan Robinson of Stafford and Denby of Lincoln, tried, unsuccessfully, to introduce 60 tonne, 25m long 'doubles' outfits like this Scania.

*Above*: As well as their Actros, Mercedes offered the Axor seen here, aimed at distribution fleets like supermakets.

*Left*: When this Foden Alpha A3 8R joined the Heygates Flour fleet in 2003 the famous Sandbach maker had just three years to go.

*Below left*: Bassetts of Tittensor are one of the UK's best-known fleets and are now part of S J Bargh but still retain their own livery. This nicely liveried DAF XF of theirs dates from 2004.

*Below*: Seddon Atkinson, a subsidiary of the Italian Iveco Group was very much on borrowed time when this '240' four-wheeled fuel tanker entered service in 2004. Apart from the grille badge it was mainly Iveco.

# The fifth decade

*Top left*: By 2005 ERFs had taken on a very MAN-like appearance - this ECT artic of R G Grant from little Offley in Herts bears some resemblance to the Gregory tanker seen below. *Above*: Volvo's FH had received a facelift in 2002 with new style headlights as seen on this 2006 registered Globetrotter of H Crabtree from New Mills near Stockport.
*Left*: This MAN TGA milk tanker of Gregory Distribution photographed outside their Cullompton depot in Devon, wears a novel Milk Marque livery with a cow 'Florence'.
*Below left*: Simmonds Transport of Telford have a smart fleet of some 80 vehicles plus extensive warehousing. This Renault Premium is typical of their fleet of 20 years ago.
*Below*: By 2009 Volvo's FH had received another facelift as seen on this timber truck.

# from 2005 - 2014

*Above*: General haulage firm A&E Transport, Sheffield run a fleet of artics and rigids throughout the UK. This Scania R450 Topline is typical of their fleet in 2010.
*Top right*: Old-established Wreford's of Northampton can boast over 120 years in the business. This nicely-liveried MAN TGX joined the 80 plus strong fleet in 2011.
*Right*: Goldstar, once one of the UK's biggest container fleets, has been part of Turners of Soham since 2016. This 2012 DAF XF seen on the A421 pre-dates the takeover.
*Below*: Former Foden users Selby's Garage of Woodhouse Eaves switched to Iveco in 2013 and its 10 strong tipper fleet is now 100% Iveco, the latest models being new X-Ways.
*Below right*: This fabulous Arocs joined the superb Paul Riches fleet in Bedford in 2014.

# Lorries in Pictures 1965-2025

Foden production finally ended in 2006. After 150 years the famous Sandbach manufacturer was axed by its American owner, Paccar of Seattle. The UK was left with just one manufacturer, namely ERF which was of course owned by MAN. This A2-4R Alpha fuel tanker was brand new when photographed near Exeter in Devon in October 2006.

Once an everyday sight, Bedfords Transport of Leeds was to face closure in 2019. Bedfords ran a fleet of Iveco Stralis artics like the one shown here which dates from 2005.

J H Jones of Old Colwyn in North Wales ran some pretty impressive outfits in their excellent livery, this being one of a number of DAF XF artics dating from 2005.

*Top left*: The owners of this 2005 Renault Premium, Inta-Site of Pollingon near Goole in Yorkshire, specialised in the transport of building materials.
*Above*: A Hingley Transport of Brierley Hill are the biggest steel hauliers in the West Midlands with a fleet of 80 44-tonne tractor units and 140 trailers. Boss Ray Hingley is also a valued member of the CVRTC.
*Left*: In their latter days ERF's ECT models were little more than re-badged MANs like this artic tanker of Hargreaves.
*Below left*: As well as general haulage, John Brindley Transport of Sharnford, Leicestershire offer specialist services with self-loading cranes, low-loaders, wide loads and so on. The company was founded in 1927 and has a smart, predominantly Scania fleet.
*Below*: Another fleet to boast a superb livery is that of P&C Hamilton of Girvan, Ayrshire, the owners of this Scania R560 Topline.

R T Keedwell Ltd of Highbridge, Somerset is one of the UK's best known haulage companies with a large nework of depot throughout the country. Typical of their nicely liveried fleet is this 2007 Scania R420 Topline based at their Carlisle depot Other locations include Glasgow, Knottingley, Hull, Haydock, Pollington, York, Cardiff and Newport.

Another firm on the author's personal list of favourites is Allinson of Darlington, the owners of the superbly liveried MAN TGA with self-loading crane and 'trombone' trailer.

With UK manufacturers dwindling, one name was to sold ier on and that was Dennis. But, by 2007 it was Dennis Eagle, part of Terberg Environmental's Ros Roca division

*Top left*: One nicely turned out tipper fleet is that of J Murphy Aggregates of Pudsey, near Leeds. Not to be confused with J Murphy in London. This nicely liveried TGS is seen at their depot.
*Above*: Famous Wincanton Transport's history goes back to 1925 and it grew into the UK's largest logistics group overseeing some 4000 vehicles. It was announced this year that it has been bought by USA giant GXO.
*Left*: The end for ERF came in late 2007, this ECT unit being the last of the line. It went to Barry Dancer haulage of East Hyde, nr Luton.
*Below left*: C Butt of Northampton took on a completely new image with the fresh livery seen on this 2008 Mercedes Actros artic. Their earlier livery was all-over green.
*Below*: Yet another favourite of the author's is Lockwood of Ambergate, the owners of this Stralis artic. They also own S&R Smith of Grays in Essex.

Renault launched its new Premium Lander eight-wheeler in 2008 and this tipper of B Jay Haulage was among the first to enter service. Judging by the livery it was possibly working for Ennstone Aggregates. It was a particularly well turned out machine with a Wilcox 'Wilcolite' body. B Jay Haulage was based in Ercall, Telford.

Winter weather can add atmosphere to a picture as it does in the case of this Rawcliffe MAN heavy haulage outfit battling the elements on the A421 near Marston Moretaine in 2009.

Bulk powder specialists Lomas Distribution of Waterswallows Road, Buxton run a wonderful fleet, this 2009 Scania R560 V8 Topline being a perfect example.

*Top left*: Tom Moorhouse & Sons Ltd of Huddersfield run a nicely liveried fleet and this DAF XF was their first truck to have a 'cab wrap' livery in which the red was a shade lighter than desired.
*Above*: Having run Ivecos, Lockwood of Ambergate added MANs in 2009 like this TGX seen on the M1 at Toddington.
*Left*: Bay Freight formed the basis of the Kinaxia Group back in 2012 which has grown rapidly over the years and now encompasses 11 leading haulage firms.
*Below left*: Hino's '700' 8x4 tippers sold well from 2006 to 2012 but they could not meet Euro 5 which took effect in 2011.This example of Carnaby's of South Killingholme dates from 2009.
*Below*: One of East Anglia's leading hauliers is Bartrums of Eye in Suffolk, the owners of this smart Actros artic.

The Scots certainly know how to paint their lorries - they really go to town and all credit to them. David Murray of Carnoustie in Angus run a fabulous fleet. This 2012 DAF XF was caught on camera at Corley Services on the M6 and is an excellent example from the much-admired fleet.

Another favourite of lorry enthusiasts is W H Malcolm, the owners of this 2012 MAN TGX artic. The famous Scottish company from Linwood celebrated its centenary in 2022.

Eight-wheelers continue to play a leading role in haulage especially in the construction sector. This 2011 DAF CF is owned by Luton based Butterfield's Building Materials

*Top left*: In 2012 Suttons Tankers paid tribute to HM The Queen's Diamond Jubilee with this specially liveried Volvo FM driven by our dear departed friend Tony Brown. *Above*: Imperial Tankers, then part of the Hargreaves Group, owned this 2011 Actros 2544. Just three years later, in 2014, the fleet was taken over by Suttons Tankers. *Left*: DHL you might say is the modern day equivalent of BRS. The German Deutsche Post owned company absorbed 'Exel Logistics', the last remnants of BRS in 2005. *Below left*:CL Transport of Bridgewater run a modern fleet of 44 tonne artics. This Renault Premium was typical of their fleet in 2012. *Below*: It came as sad news to haulage enthusiasts in 2023 when Sam Anderson of the Anderson Group, one of Scotland's most admired fleets, ceased trading. This Scania R440 is a classic example of their lorries.

No one could deny that modern lorries like this 2013 Renault Premium of Damac Transport from Goole are every bit as interesting as those of the past. This nicely painted 38-tonne outfit with its Feldbinder bulk powder tank makes an impressive sight by any standards. It was photographed heading up the A14.

Also seen on the A14, but near Hemmingford Abbotts in Cambridgeshire, is this smart Actros MP4 of Newcastle-on-Tyne based Murray Hogg, in their Stelrad contract livery.

Taken at the same spot is this 2013 DAF XF in the fabul- ous livery of another Newcastle-on-Tyne firm, J R Adams who began operations back in the early 1930s.

*Top left*: Another fleet with an eye-catching livery is that of A J A Smith of Clitheroe in Lancs. They narrowly escaped closure in 2009 but were soon back up and running. This MAN TGX dates from 2014.
*Above*: In recent years Mercedes with their Arocs eight-wheeler have proved popular in the construction world as tippers and mixers.
*Left*: One firm, based at Alnwick in Northumberland, namely Manners Transport, boasts one of the best liveries of all. This XF Super Spacecab low-loader is seen on the A507.
*Below left*: One of the biggest and best-known East Anglian fleets is that of Wimblington based Knowles Transport who began back in 1932. The firm runs 200 units and 350 trailers in the striking livery seen on this 2013 FH.
*Below*: Another well turned out outfit seen on the A14 is this 2014 Renault Range T of Weaver Haulage from Ashbourne, Derbs.

# The sixth decade

*Top left*: In recent years the Gregory Group has expanded with major acquistions including A R R Craib, Pollock, Hayton Coulthard and nov John Mitchell, Grangemouth. With 2 million sq. of warehousing, 1200 tractors and 2000 trailers i is among the UK's biggest distribution companie
*Above*: Craig's of Denny, the owners of this 201 Renault Premium seen on the A1(M), have one of the best liveries in the business.
*Left*: Another fabulous fleet is that of Tyson H Burridge of Distington who operate throughout the UK. This DAF XF dates from 2017.
*Below left*: Tippers working for the large Aggregate Industries Group are mainly owned by franchisee hauliers.
*Below*: Ed Weetman from Great Haywood near Stafford run a 30 strong fleet in this striking and rather patriotic livery, as on this 2019 Scania.

# from 2015 - 2024/5

*Above*: Harlow Timber, of Long Whatton, once ran Fodens but switched to Iveco when Foden ended. The old established firm boasts 13 branches and an online ordering service.
*Top Right*: With a fleet approaching 2000 trucks Maritime Transport of Felixtowe is one of the UK's largest container haulage specialists. Maritime runs a mix of Volvos, Scanias, DAFs MANs and Mercedes, like this Actros 2545.
*Right*: Heygates is one of the UK's biggest millers based at Bugbrooke, Northants with 3 other mills including Downham Market. It majors on DAFs like this 2022 CF artic bulk tanker.
*Below*: One of the UK's leading bulk powder fleets is N S Clarke of Rugeley who own this 2023 Volvo FH seen in Flitwick town centre.
*Below right*: Famous Eddie Stobart became part of Muller-owned Culina Group in 2021.

Maritime Transport, the UK's largest logistics company, majors on container traffic with impressive outfits like this 7-axle 'combi', headed by a 2015 Volvo FH. Maritime's boss, John Williams, is the son of Tommy Williams of South Wales based TDW Distribution. A decade or so after this shot was taken on the A507, Ampthill-Flitwick bypass, the flee became part of Swiss-owned Medlog which is the logistics arm of the Geneva based Mediterranean Shipping Company.

Since the early '70s Bring Cargo, the owners of this 2015 Renault Range T, have been operating logistic services and are based in Stallinborough near Immingham, Lincolnshire.

Woodrush Ltd, the name on this 2015 Volvo FE, were a well established haulage company and Pallex member based in Mid-Beds until they closed about 3 years ago.

*Top left*: WS Transportation was founded in 2014 by William Stobart of Eddie Stobart. Their HQ is in Warrington but they operate nationwide mainly for the construction industry. This Scania 8x2 dates from 2016.
*Above*: One of Scotland's nicest liveried fleets is that of James Hislop from Kelso.
*Left*: A real surprise was this one-off prototype 44-tonne Dennis Eagle 'Elite 4' artic for Explore, seen on test in 2017.
*Below left*: Fermac's were a familiar sight up until 2020. That year they merged with Agribulk to form Rockley Transport who specialise in bulk tipping work. The firm now has 40 depots around the UK and its HQ is in Sudbury, Suffolk.
*Below*: One well-known firm, TWT from Ystrad Mynach in South Wales, sadly shut down in 2024 after 44 years in business. This 2015 Actros is typical of their fleet.

Allinson's of Darlington, who own this Actros, is one of the best-known and best-liveried fleets in the North East. It was founded by George Allinson in 1955 and, 70 years on, it operates 120 vehicles and 250 trailers. It boasts 250,000sq.ft of warehousing and operates nationwide with other facilities in Yorkshire, Cambridgeshire, Gloucestershire and Liverpool.

Another favourite haulier is Longs of Leeds, the owner of this smart DAF XF. Founded in 1918, it now operates a fleet of 100 trucks and is a member of the Pall-ex network.

Also from Yorkshire is this Rhodes Scania. Rhodes of Yor transports Portakabins. Such work requires a large lorry-mounted crane, hence the need for an 8-wheeled unit.

*Top left*: R D Williams Transport's normal livery is pale yellow and green so this 2018 Iveco in silver and black, seen on the A421, was something of a surprise.
*Above*: Now one of Bedfordshire's biggest hauliers, D&G Noble specialise in self-loading and offloading vehicles like this Scania R500 eight-wheeler.
*Left*: Founded back in 1931, H R Maiden of Telford run a nicely-liveried modern fleet of artics like this 2018 registered MAN TGX. Since 2017 the firm has been part of Kinaxia.
*Below left*: H Crabtree Ltd of New Mills in Derbyshire run a particularly smart fleet. Founded in 1961, the firm is a member of the Pallet-Track network.
*Below*: In recent years there has been pressure, mainly from Transport for London, to develop low-profile, cyclist-friendly construction vehicles like this Mercedes Econic mixer.

Bulk powder tankers are among the most impressive outfits on the road especially the likes of this Scania S500 Topline of Lomas Distribution of Buxton. LD19OAR is seen here with a Feldbinder tank on the A507, most likely heading for Weber St Gobain's plant on Flitwick industrial estate.

Driffield based Robinson Contract Services' fleet of bulk tippers are solely engaged on transporting 'Yorgrow' biosolids and sewage sludge fertilizer for Yorkshire Water.

Shirley's of Cellarhead's fabulous fleet includes a numbe of Scania R450 XT eight-wheeled tankers with drawbars c Intermol molasses who now trade as parent ED&F Man.

*Top left*: One of East Anglia's best-known tipper fleets is G Webb of Cambridge. With a fleet of 65 they've been going for 78 years.
*Above*: A familiar sight in many parts of the UK, Marshalls DAFs, like this outfit seen on the A421, deliver block paving. Based in Huddersfield, the firm has several locations including Elstow, near Bedford.
*Left*: Mason Bros of Swineshead near Boston, Lincolnshire operate a much admired fleet mainly engaged in agricutural produce. This 2020 DAF XF is seen on the A50.
*Below left*: Recently, Gemmix lorries like this 8cu.m DAF CF have become a familiar sight in East Anglia. Gemmix is next door to the Cemex plant in Kempston Hardwick.
*Below*: George Davies Turf, based in Olney, Buckinghamshire, supply turf over a wide area, as far as London and Oxford using modern Scania eight-wheelers like this curtainsider.

The unmistakable bright yellow lorries of Jack Richards & Son are to be seen all over the UK. Their mainly DAF fleet is a great favourite among enthusiasts. Since 2019 the firm has been part of Turners of Soham but has retained its distinctive livery. Established back in 1956, the company once had a large collection of classic vehicles. This XF is seen on the A17.

Iveco's new X-way models were hot off the press in 2021. This 6x2 rigid with self-loading crane of MKM building supplies was new to their Milton Keynes branch that year.

Bulk hauliers Alfred Hymas, the owners of this Volvo FH, were founded in Burton Leonard in 1911. They now run the biggest and best known tipper fleet in North Yorkshire.

*Top left*: GE-BE of Swaffham and Kings Lynn have been going over 40 years having begun in 1982. Their 70 strong fleet wears this bold red livery with gold 'GE-BE' logo.
*Above*: Another favourite among enthusiasts is bulk transport firm B J Waters of Matlock who have traded since 1927. This Scania 500 S is typical of their superb modern fleet.
*Left*: Another fleet to boast a fabulous livery is ITS, aka P A Donaldson, of Irthlingborough. Founded in 1979 by Peter Donaldson, ITS operate around a dozen trucks and own a few classics too including a Kenworth K100.
*Below left*: Goldstar Heathrow of Slough runs an extensive fleet from vans to 44 tonne artics mainly on airfreight deliveries, including this New Generation DAF XF.
*Below*: H Nicholls of Sittingbourne run an all-Iveco fleet of some 100 trucks and 350 trailers including many new S-Ways like this outfit seen on the M1 near J12.

# Lorries in Pictures 1965-2025

The name Freightliner dates back to the '60s but the present day organisation of that name is very different and is internationally based. Originally a division of British Rail responsible for handling shipping containers it was later privatised in the 1990s. It is now a large intermodal transport network with around 300 trucks. Many are of the 'combi' type as in the case of this 2023 registered Mercedes Actros 2551 with 'Gigaspace' cab seen on the A421 near Marston Moretaine.

Freshlinc, part of Lincolnshire Field Products, runs a large fleet of temperature-controlled artics from its base in Spalding. Most are modern Scanias like this 500 S 'Topline'.

In the lighter 4x2 category, this Volvo FL 18-tonne gross curtainsider fills a role. At one time such lorries might have been 16-tonners from Leyland, Ford or Bedford.

## Lorries in Pictures 1965-2025

*Top left*: Since its launch in 2021, Iveco's S-Way has found its way into a lot of high profile fleets including Sparks of Glastonbury, the owners of this smart outfit. *Above*: In recent years there has been a shift from conventional 8x4s to 'Tridems' of this type, seen with Jet Plant Hire road planing. *Left*: Another type of rigid-eight that has gained in popularity is the 8x2 with single-wheeled tag axle. This Arocs of Builders' Merchants E H Smith is a perfect example. *Below left*: DJV Transport & Forwarding of Tilbury run a modern fleet including 44 tonne artics like this MAN TGX. They have depots in Medway, Newport, Hull and Blackburn. *Below*: A familiar sight around East Anglia are the nicely liveried bulk tippers of P J Thory of Coates, Peterborough who are leading suppliers of sand and gravel. Their fleet includes DAF rigid-eights and artics.

# Lorries in Pictures 1965-2025

Volvo announced its new FH Aero range in January 2024 claiming its improved aerodynamics reduces fuel consumption and emissions. It is also available in electric form, following a growing trend to switch away from 'fossil' fuels. This is one of a number of FH Aeros ordered by Chiltern Distribution of Peterborough who have specialised in temperature controlled distribution of food and pharmaceutical products throughout the UK and Europe, over the past thirty years.

Previously known as TNT Post UK, Whistl Group, the owners of this Renault Range T, are a logistics company that specialises in mail, parcels and fulfillment for ecommerce.

R D Williams & Sons of Easton near Huntingdon are said to be the first UK customer for the new Iveco S-Way launched in 2021, the year of their 50th anniversary.

*Top left*: A new addition to the Marshalls fleet is this DAF XD-C 410 rigid eight bulker to deliver mortars and screed mix.
*Above*: Explore Plant & Transport was established in 2015 when WS Group took over Laing O'Rouke at Worksop. They now own 7 depots and 150 trucks, mainly Scanias like this 500 R 8x2.
*Left*: With 250 vehicles and 8 UK depots, Clarke are among the UK's biggest family run logistics companies. This Actros 18 tonne 4-wheeler is one of 35 trucks based at their Luton depot in Barton-le-Cley.
*Below left*: GBRS stands for 'Growing Beds Recycling Service'. Based in Bedford since 2005, it specialises in organic waste recycling. This FH is a recent addition to the fleet.
*Below*: When talking trucks we musn't forget the all-important lighter models like this DAF XB-190, 7.5 tonner owned by gasket and seal manufacturer Avon Group based in Bristol.

What better way could there be to round off the pictures than with a brand new eight-wheeled curtainsider? This DAF XF delivered to Bedfordshire haulier D&G Noble of Wootton ready for March 1st 2025 is a truly magnificent machine.

D & G NOBLE
HAULAGE
DAF
KU25 YKC
NOBLE
Haulage
www.dgnoble.c
D & G N
Haulage

# Meet the Committee

The CVRTC is run by a committee of volunteers, all of whom have strong connections to the transport industry. Here are some brief outlines of current committee members and club officials:

**Graham Edge** edits our regular quarterly magazine 'CVRTC News' as well as being responsible for membership renewals. Graham has been a club member for 39 years and has held a lifetime interest in road transport. In that time he has been involved in lorry restoration, notably of a rare ex-BRS Seddon, FRC805C, and an AEC Mandator artic in Spiers of Melksham's livery, TGU205M. He is also a noted transport writer and author and has produced numerous books. He still works full time in the industry as an operations manager for Turners (Soham) Ltd.

**Malcolm Mortimer** is CVRTC chairman as well as being organiser of the club's winter meetings at Coventry for the past 35 years. Malcolm has held a keen interest in road haulage since 1957 and spent his working life in the transport industry. He is also responsible for recruiting and deploying marshalls at the annual Gaydon show. With his extensive knowledge of haulage history and present-day developments he contributes most of the information published in the CVRTC News 'industry news' pages. As well as this, Malcolm is a keen photographer with a large photographic archive.

**Stephen Smith** is CVRTC secretary (a post formerly held for decades by Steve Wimbush) and has been fascinated by all forms of transport from a very early age. He enjoyed a 45-year managerial career with the Birmingham & Midland Motor Omnibus Co (better known as 'Midland Red') and its successors before finally retiring at the end of 2016, aged 62. Steve has been a keen transport historian and photographer since the late 1960s and has owned several classic lorries including a couple of LAD cabbed Leylands.

**Richard Stanier** has been a CVRTC member for over thirty years and has the responsibility of being the club's social media administrator as well as recently taking on the job of club treasurer. He is actively involved in vehicle preservation, owning a number of tractor units including a Guy Big J4T, B101SFK, and a Seddon Atkinson 401, D21OTH. Richard has collected a huge archive of commercial vehicle literature and is an accomplished photographer as well being a regular contributor to industry trade journals.

Husband and wife **Martin and Donna Platt** play an important role as club ambassadors, attending shows and events around the UK with the CVRTC outdoor stand, greeting members and generally spreading the word for the club. Martin is active in lorry preservation and is best known for his 1971 ex-Robsons Foden S39 unit WHH241J, Fleet No.FD.513, 'Border Banshee' but he also has an AEC Ergo Mandator and an ex-Bassett's Foden unit. Martin works as a master technician at Sapphire Vehicle Services carrying out truck maintenance and MOT preparation for Argos and DHL.

# About the Author

Peter Davies has held an all-consuming interest in lorries his whole life. He has been involved in every aspect of the subject having served thirty years in the marketing department of Bedford Trucks and since the closure of the famous manufacturer in 1986, has worked as a freelance transport journalist, photographer and technical advisor. He held a Class One HGV licence for over 30 years and has driven a huge variety of heavy trucks, both early and modern. He has contributed to leading magazines over the years including *Truck*, *Truck & Driver*, *Trucking International*, *Classic Truck*, *Classic & Vintage Commercial* and others. He has produced a dozen books including a World Truck Encyclopedia.

He is a keen model collector and worked as freelance designer draughtsman and advisor to Corgi Classics from 1996 onwards and was responsible for introducing Corgi to haulage liveries and accurate cabs and loads. He designed and introduced completely new Corgi models such as short wheelbase 8w tippers, Chinese sixes, artic bulk powder tankers, curtainsider trailers and life-like roped and sheeted loads, all of which were new to the Corgi management. All in all, he introduced around 28 new models including French trucks like the Berliet GLR and the Renault Fainéant. He provided Corgi with colour pictures of famous fleets to ensure accuracy of liveries.

While heavily involved in masterminding lorries in miniature, Peter also collected the real thing and is best known for his Maudslay Meritor and AEC Mammoth Major Mk.III eight-wheelers. His collection peaked in the 1980s with a Sentinel, several Leylands, a Maudslay Mogul, an Atkinson Borderer and an Albion Chieftain. He also shared a Bristol HA6G and an ERF KV eight-wheeler.

Peter has always enjoyed doing illustrations of lorries and, in 1981, he was commissioned by the NFC to illustrate a calendar on the early years of BRS. This became a collector's item and the NFC were so pleased with it that Peter was asked to repeat the exercise in 1985 with illustrations of famous BRS Contract liveries.

When the idea of a club for lorry enthusiasts was first proposed in 1965 Peter was among the founder members and has played a role in CVRTC affairs from the very beginning. He was also, with support from other members and his wife Jill, responsible for creating what became one of the UK's major lorry shows, namely the Classic Commercial Motor Show, launched at the Cranfield LDoY Finals in 1979. Peter has served as CVRTC President for over thirty years.

Made in the USA
Monee, IL
02 November 2026

67381625R00050